NATURE'S GROSSEST

SEAGULLS EAT TRASH!

By Natalie Humphrey

Please visit our website, www.garethstevens.com. For a free color catalog of all our high-quality books, call toll free 1-800-542-2595 or fax 1-877-542-2596.

Library of Congress Cataloging-in-Publication Data
Names: Humphrey, Natalie, author.
Title: Seagulls eat trash! / Natalie Humphrey.
Description: Buffalo, New York : Gareth Stevens Publishing, [2024] | Series: Nature's grossest | Includes index.
Identifiers: LCCN 2022056734 (print) | LCCN 2022056735 (ebook) | ISBN 9781538285787 (library binding) | ISBN 9781538285770 (paperback) | ISBN 9781538285794 (ebook)
Subjects: LCSH: Gulls–Juvenile literature.
Classification: LCC QL696.C46 H86 2024 (print) | LCC QL696.C46 (ebook) | DDC 598.3/38–dc23/eng/20221129
LC record available at https://lccn.loc.gov/2022056734
LC ebook record available at https://lccn.loc.gov/2022056735

Published in 2024 by
Gareth Stevens Publishing
2544 Clinton Street
Buffalo, NY 14224

Designer: Leslie Taylor
Editor: Natalie Humphrey

Photo credits: Series art (background) Oleksii Natykach/Shutterstock.com; cover Emagnetic/Shutterstock.com; p. 5 Rosamar/Shutterstock.com; p. 7 kipgodi/Shutterstock.com; p. 9 FJAH/Shutterstock.com; p. 11 Daniil Skoblov/Shutterstock.com; p. 13 Leandro Felipe Santiago/Shutterstock.com; p. 15 BalkansCat/Shutterstock.com; p. 17 Ruben CL/Shutterstock.com; p. 19 Manuela Durson/Shutterstock.com; p. 21 JADA photos/Shutterstock.com.

Printed in the United States of America

CPSIA compliance information: Batch #CSGS24: For further information contact Gareth Stevens at 1-800-542-2595.

CONTENTS

Boldface words appear in the glossary.

Gross Birds

Seagulls are found around the world! There are many kinds of seagulls. Each kind of seagull looks and acts a little bit different. But all seagulls have one gross thing in common: they love trash!

Kinds of Seagulls

There are over 50 **species** of gulls in North America, all commonly called seagulls. These birds are found in many kinds of homes. Some live close to water. Others can be found anywhere there's food to eat.

Garbage Dump Home

Seagulls aren't just found in the wild though. Seagulls love places with a lot of humans. They can be found in parking lots, parks, and even garbage dumps! Some people even consider seagulls **pests**.

Groups of Gulls

Gulls sometimes live in large groups called colonies. Seagulls like large, flat spaces where they can lay their eggs and find food. Seagulls usually have their nests in the same area they hunt for food.

Hungry Seagulls

Seagulls will try to eat nearly anything! In the wild, they eat food left by other animals. But seagulls will also eat anything humans throw away. Some seagulls will even steal food right out of your hand!

Trash Eaters

Seagulls don't just eat human food. Seagulls have strong stomachs and eat things that aren't food at all! Scientists have found a lot of strange things in seagull stomachs. Seagulls eat building **materials**, paper, and even plastic.

Extra Gross!

Eating trash isn't the only gross thing that seagulls do. When a seagull eats something they shouldn't, they'll throw up a pellet. A pellet is a clump of food that a bird can't **digest**. Pellets can be made of anything!

Dangerous Poop

Seagulls eating trash is gross, but it can also be **dangerous**! Because seagulls will eat so many things, sometimes their poop can cause **pollution**. When seagulls poop in water, they may cause pollution in the water.

Moving Day

Seagulls' love for trash has also changed where they decide to live! Seagulls will **thrive** anywhere there is enough food. When humans make more trash, seagulls have more food to eat. The more food there is, the closer seagulls move to us!

GLOSSARY

dangerous: Unsafe.

digest: To break down food inside the body so that the body can use it.

material: Matter from which something is made.

pest: An animal that causes harm to humans.

pollution: The act of water or land becoming unsafe for people or animals.

species: A group of plants or animals that are all of the same kind.

thrive: To grow and live well.

FOR MORE INFORMATION

BOOKS

Rae, Susie. *What's the Difference?: Animals.* New York, NY: DK Publishing, 2022.

Montgomery, Sy. *The Seagull and the Sea Captain.* New York, NY: Simon & Schuster Books for Young Readers, 2022.

WEBSITES

Britannica Kids
kids.britannica.com/kids/article/gull/353219
Find out more fun facts about seagulls.

DK Find Out
www.dkfindout.com/us/animals-and-nature/birds/gulls/
Learn more interesting facts about gulls and find out how to spot them in the wild.

Publisher's note to educators and parents: Our editors have carefully reviewed these websites to ensure that they are suitable for students. Many websites change frequently, however, and we cannot guarantee that a site's future contents will continue to meet our high standards of quality and educational value. Be advised that students should be closely supervised whenever they access the Internet.

INDEX